A complete steam-powered electricity-generating set was a child's dream. This set by Bing from about 1910 comprises a vertical steam engine, dynamo and period lamp. The engine speed is stepped up to about 4000 rpm to operate the dynamo, which produces about 4 volts. (300 by 150 mm; height 320 mm.)

TOY STEAM ACCESSORIES

Marcus Rooks

Shire Publications Ltd

CONTENTS

Published in 1996 by Shire Publications Ltd, Cromwell House, Church Street, Princes Risborough, Buckinghamshire HP27 9AA, UK. Copyright © 1996 by Marcus Rooks. First published 1996. Shire Album 324. ISBN 0 7478 0313 7.
Marcus Rooks is hereby identified as the author of this work in accordance with Section 77 of the Copyright, Designs and Patents Act 1988.

Printed in Great Britain by CIT Printing Services, Press Buildings, Merlins Bridge, Haverfordwest, Dyfed SA61 1XF.

British Library Cataloguing in Publication Data: Rooks, Marcus. Toy Steam Accessories. – (Shire Album; 324) 1. Toys 2. Mechanical Toys 3. Steam I. Title 688.7'28 ISBN 0-7478-0313-7

ACKNOWLEDGEMENTS
The author would like to acknowledge, with thanks, the help of Simon Ashington and Esmie de la Cruz in the preparation of this manuscript; Mike Cooke for his support and advice; and Bob Gordon for kindly permitting the use of the line drawings by Ann Speedy of the manufacturers' trademarks on pages 29-31. All the examples illustrated are from the author's own collection with the exception of the following: Mike Cooke, pages 5 (top), 21 (bottom), 22 (top), 25 (bottom), 29 (top); Mike Green, page 23 (top left); Tim Taylor, pages 19 (top), 20 (bottom), 21 (centre), 26 (both).
Photography, including the front cover, is by Brian Woods.

Cover: *This delightful windmill, complete with its occupants, was made by Wilhelm Kraus about 1930. When in operation the sails turn and the donkey, with its loaded sacks, circles in and out of the mill, followed by its master. The mill is in lithographed tinplate and the base is hand-painted.*

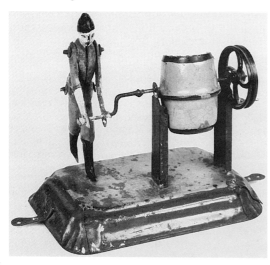

This is a very early (around 1900) German hand-finished figure operating a butter churner. The figure is articulated at the waist to allow a bending movement when in operation. These early hand-finished items were painted after final assembly and tend to suffer more paint damage than later lithographed items. (110 by 80 mm; height 110 mm.)

This illustration from the box lid of a steam toy by Plank shows two boys happily playing with a large vertical steam engine which is coupled via a countershaft to some accessories. One is a lathe in a rather crude workshop. The other is a pillar drill of almost full-sized proportions. All dates.

INTRODUCTION

Ever since the introduction of the first toy steam engines during Victorian times people have asked what they could do. Although there is no doubt that the steaming of these stationary engines held a great fascination for the Victorian boy, once they were up and running they had very little play value. This was a great disadvantage when they were compared with steam locomotives, boats and traction engines, which at least moved.

The major toy manufacturers, especially the great German ones, were quick to recognise this. In order to maintain sales of their stationary engines, they started to produce a huge variety of accessories for them to drive. As the name suggests, accessories were separate toys that could be coupled to a steam engine that operated them. In many ways these accessories were more colourful and interesting than the engines themselves, and to the collector they present a treasury of desirable

items. The range of accessories was enormous, seemingly limited only by the imagination of the designers. They ranged from the simplest hammer to a complete workshop, from single figures to entire fairgrounds. Some even came with their own musical-box accompaniment.

Most of the accessories were made by the great German firms of Bing, Doll, Marklin and Carette, although others, such as Mohr, Hess, Falk and Fleischmann, made significant contributions. English firms such as Mamod and Bowman made a limited range only, restricted to machine tools, and these were dull compared to the German-made accessories. The accessories made by Marklin are particularly highly prized because of their reputation as makers of high-class toys.

Most accessories date from before the Second World War, although a few firms continued the tradition afterwards, but at a greatly reduced volume and variety.

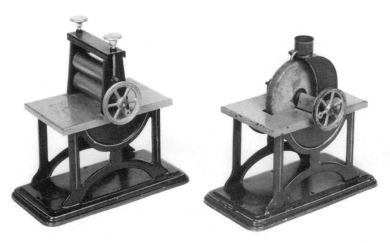

These accessories by Doll from about 1930 make use of the same pressing for their base. One is a large-scale grindstone, complete with water reservoir. The other is a washing mangle, for the use of the daughters of the family. (130 by 65 mm; height 130 mm.)

HISTORY OF MANUFACTURING

Toy steam engines started to appear in the middle of the nineteenth century. The first ones were made by specialist optical firms that had branched out into toymaking. It was not until twenty years later that large-scale commercial production of steam toys and accessories began, mainly in Germany. This industry was largely based in Nuremberg, which was the centre of the German commercial tinplate industry. The famous firms of Ernst Plank (1866), Peter Doll (1878), George Carette (1886) and Bing Brothers (1866) were all situated in Nuremberg, although Marklin, perhaps the greatest of them all, was in Göppingen. By 1900 the German firms had established themselves as the leading toymakers in the world.

Accessories were made from tinplate, brightly coloured and imaginative, but

A delightful water toy by Bing from about 1930, rarely found in such pristine condition. This example is hexagonal and hand-finished in cream and gilt. The operating mechanism is clearly visible, comprising a brass oscillating pump connected to the drive shaft. Unfortunately the fountain of water is no more than a dribble. (Diameter 170 mm; height 70 mm.)

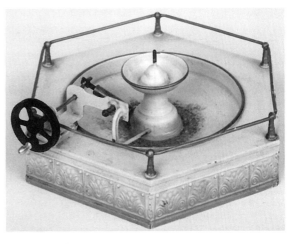

4

A collection of contemporary accessories by the English firm of Mamod, powered by a Mamod steam engine. All these tools are available today, but they have been in production, virtually unchanged, since they were first introduced in the 1930s.

they were rather poorly constructed. Initially they were made entirely by hand. The individual pieces of tinplate were cut, stamped and bent, then soldered together and finally painted. The process was extremely laborious, slow and expensive.

As demand steadily increased, new techniques were introduced to speed up production. Around 1900 the introduction of chromolithography radically altered the way in which steam accessories were produced. Chromolithography is the process by which sheets of flat tinplate can be printed in colour with the finished design. This is done in a number of stages, each one printing a different colour until the

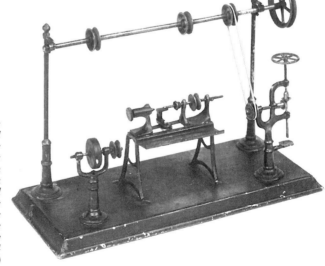

A typical example of the work stations produced by all the major toy manufacturers. This one is by Carette from 1910 and comprises a lathe, drill and grindstone. These are of a very delicate construction and are for effect only. (200 by 90 mm; height 104 mm.)

An unusual rotative toy by Plank, in the form of a simple fan. It is made from hand-finished tinplate, the maker's mark from about 1920 being clearly visible. (Diameter 100 mm; height 180 mm.)

final design is achieved. In this way elaborate and highly detailed pictures could be produced – something that hand painting could not achieve. Carette was especially renowned for the quality of its lithography.

The printed sheets still needed to be cut, stamped and bent to shape. However, as any heat, such as that applied during soldering, would damage the printing, a new method of assembly was needed and the now familiar method of slot and tab construction was adopted. A tab on one end of the piece was inserted into a slot on the other and bent over, thus holding the two edges firmly together. Pulleys and

other heavy parts were usually cast in a cheap lead-based alloy, using simple metal moulds. In this way the casting needed very little finishing. Although the final assembly was still done by hand, machine production methods were used, thereby increasing the output and improving the quality.

During the First World War toy production virtually ceased. After the war there were shortages of labour and materials so that it was not until 1920 that production approached pre-war levels. But the commercial world had changed. In Britain, especially, there was pressure to buy home products and this sentiment allowed a number of British toymaking firms, such as Bowman, Wormar, Hobbies, Mersey Models and Barr Knight, to become

This scene by Marklin makes use of a water pump hidden in the millhouse to circulate water from the lake to the horse trough and watermill. Made in hand-finished tinplate, from about 1930. (150 by 150 mm.)

6

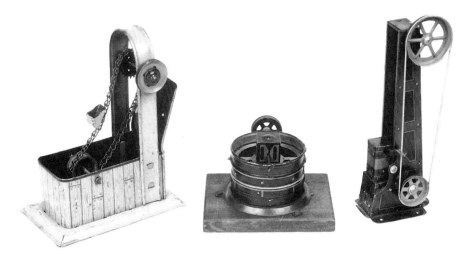

A collection of accessories by Hess, all from about 1925. (Left) A sand dredger, which could easily by mistaken for a water toy, but close inspection reveals that the hopper is not watertight and is suited only for materials such as sand. (Centre) A simple butter churner. The body conceals the two paddles which produce the butter. (Right) A trunk elevator. The trunk is lifted to the top of the rails by the spring belt. On reaching the top it is released to fall back with a satisfying crash.

Wilhelm Kraus made this attractive millhouse in about 1930. The detailed lithography shows the washing on the line and stones to keep the roof in place, which was typical of rural German life. The waterwheel is operating a simple trip hammer. A number of similar accessories were built using the same pressing. (180 by 120 mm; height 110 mm.)

7

This version of the smithy is by Arnold, about 1920. The anvil conceals a friction-drive motor driving an abrasive wheel. The work contains a flint, which, when struck by the hammer against the wheel, produces a shower of sparks. (120 by 80 mm; height 120 mm.)

This smithy is by Fleischmann, about 1950, and has been given a crude open shelter. The operating mechanism is hidden from view; its action is simply to raise and lower the smith's arm. There are no sparks from this version. (140 by 110 mm; height 110 mm.)

established. It is unfortunate that they concentrated their efforts on making steam engines. None of them made accessories to any significant extent. Bowman, Mersey Models and later Mamod made a limited range of machine tools. Although well-made, these could not compete with the German accessories. The German firms may have lost much of the trade in steam engines but they still held a virtual monopoly in the manufacture of accessories.

This very early hand-finished man at a sawbench is by Mohr & Kraus, about 1906. So dangerous a working practice would alarm a modern safety inspector. (113 by 76 mm; height 123 mm.)

This simple German fountain of about 1930 is made from hand-finished tinplate. The fishes are somewhat later and are made from plastic. (Diameter 170 mm; height 90 mm.)

During the 1930s the general economic downturn and the depression affected the toy industry and a number of the German toymakers, such as Bing, Doll and Plank, ceased production.

The Second World War, like the First, had a dramatic effect on the toymakers. By 1940 all toymaking had ceased and it was not until 1949 that production resumed, and even then on a vastly reduced scale. In Germany only Fleischmann were producing accessories, ceasing in the

9

1960s. However, at that time the firm of Wilesco started to produce a range of novel and interesting accessories. In the 1990s they alone are producing accessories in any quantity and have started to manufacture some of the working men that so characterised pre-war accessories.

In Britain only Mamod continued the accessory tradition and they are still in production today. In the 1950s SEL appeared on the scene, making a range of rather small engines and a series of machine tools. Their toys were characterised by the use of thermosetting plastic for the bodies, but as they were to be used in the presence of lighted spirit and steam their usefulness was limited. They continued in production until the mid 1960s.

In the United States of America the firm of Weeden tried to break into the market and from the 1880s until the 1920s they sold a range of machine tools and a windmill. Flimsily constructed and painted a rather dull orange and green, these toys were never very popular and they are rarely found today.

One hundred years after the introduction of accessories, the market for them, as well as for the steam engine, is virtually dead. In the 1990s it is computers and electronic games, and not steam toys, that hold sway.

Doll made this unusual horse trough from hand-finished tinplate in about 1910. The trough is made to simulate a hollowed-out log. The central column supports the drive shaft and conceals a pump which circulates the water through the two downpipes. (140 by 70 mm; height 160 mm.)

A typical layout of horizontal trip hammers, possibly by Schoenner, 1910. As the shaft is rotated, cams trip the hammers in sequence. As the cams are usually set at an angle to each other, a continuous barrage of noise is produced. The body is made from sturdy castings and the base is wood. (120 by 90 mm; height 60 mm.)

ACCESSORIES

Stationary steam engines usually possessed a large flywheel and a smaller driving pulley. The accessory would have its own driving pulley connected to the operating mechanism. When in use these pulleys would be coupled together, using a spring belt or similar, and thus the motion of the engine was transmitted to the toy.

Accessories by different makers could be coupled to the same engine. The pulley of the accessory usually also possessed a handle, allowing it to be operated by hand if no engine was available.

The operating mechanisms were usually extremely simple. In rotative toys such as windmills the pulley was connected directly to the shaft turning the sails. Sometimes a set of gears was in-

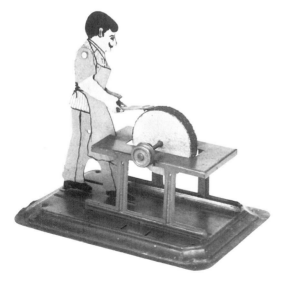

Left: *Most manufacturers produced a figure at the grindstone. This one is by Arnold, about 1930. The flat figure is made from lithographed tinplate. The action of the emery-covered wheel against a flint produces a shower of sparks. (120 by 80 mm; height 120 mm.)*

Left: *Plank made this rather flimsy example of twin vertical drop hammers about 1920. The hammers are lifted by the operating crankshaft and are allowed to fall under their own weight. The maker's trade mark is clearly visible, almost dwarfing the accessory. (80 by 50 mm; height 80 mm.)*

Right: *A simple but well-made pedestal grindstone by Bing from about 1930. The body is made from diecast metal. The wheel is made of wood and is for effect only. (90 by 90 mm; height 100 mm.)*

terposed and these could either increase or decrease the speed of rotation. With figures, usually the only part that moved was the arms. These would be operated by a simple crankshaft, with operating levers connected to them. In toys such as fountains the pulley would drive a simple oscillating pump, which would provide the necessary jet of water. With machine tools the mechanism would depend on the type of tool. Some of these possessed complex gearing and eccentrics to provide the necessary movement.

MACHINE TOOLS

Machine tools were the staple accessories of the stationary engine. These miniatures mirrored the design of their full-

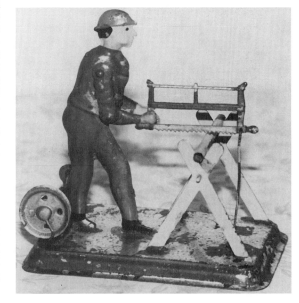

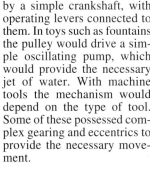

This working man was probably made by Carette in about 1910. Being made and painted totally by hand, it shows a considerable amount of paint loss from the base.

This well-made drill press with a one-piece diecast body was made by Bing in about 1930. The drawback with this design is that there is no provision for the rise and fall of the table, so it cannot be used for the purpose intended. (90 by 90 mm; height 110 mm.)

Above: *This sturdy vertical reciprocating saw was made by Doll about 1920. The body is made from diecast metal. The four blades are activated by a crankshaft and would have been used to produce planks. Doll made another version using pressed tinplate for the body, at the same period. (80 by 80 mm; height 130 mm.)*

Above: *This sturdy woodworking lathe by Plank is an accurate copy of full-sized lathes of the period around 1910.It has a headstock, tailstock and a toolpost rest. (130 by 74 mm; height 110 mm.)*

Right: *A modern circular saw by Wilesco, 1975. It conforms to modern safety standards in that it incorporates a safety guard.*

One of the many work stations made by Plank. The overhead lineshaft, supported by finely cast supports, has provision for eight accessories. This example of about 1900 has a lathe, grindstone, saw and trip hammers. It has a wooden base covered with printed paper simulating a tiled floor. (500 by 120 mm; height 230 mm.)

sized counterparts, thus giving an unusual insight into the design of what were then modern machine tools.

The *hammer* was perhaps the most popular tool, being found in a wide variety of forms. Hammers were popular with the manufacturers because they were cheap and easy to make, and they were popular with the consumers because they could make a tremendous noise that would delight their young operators.

The *grindstone* was also well represented. This was easy to manufacture; the body could be either a casting or a tinplate pressing. As in full-sized practice, there were many variations, single- or double-headed, pillar- or bench-mounted. Some came with their own water reservoir, for cooling or dust suppression. Most miniature grindwheels were made from wood and could not be used for the purpose intended. Some, however, possessed an abrasive rim that could actually be used, and the larger ones were capable of grinding wood and soft metals.

Drill presses are frequently encountered. These varied from the simplest pillar drill, which was for effect only, to more elaborate ones which possessed working drill chucks and quills. These were as much as 20 cm (8 inches) high and bordered on full-sized working machines.

Sawbenches formed another large group

of tools and had every conceivable arrangement of cutting blade, from simple circular saws to reciprocating hacksaws. Again the smaller toy versions were for effect only, but the larger ones were capable of cutting soft woods such as balsa.

The *lathe*, that indispensible machine tool, was also reproduced in miniature, although not so frequently. Like their full-sized counterparts, these toys were sturdily built using cast metal parts which were quite accurately detailed. Both metalworking and woodworking lathes were produced. Marklin and Carette produced lathes with working chucks and simple tooling, making simple turning operations possible.

Marklin, especially, produced an extensive range of specialist machine tools, all produced to a very high standard. Boring, shaping and planing machines were the most frequently produced.

Most manufacturers grouped a variety of their machine tools into a simple work station, some even including a steam engine. They were usually for effect only but are fascinating to watch in action. Thus the budding engineer could be the foreman as well as the operator.

WATER TOYS

Water toys have always been popular; anything that involves water splashing everywhere has an obvious appeal to

A Fleischmann water toy of about 1950, which has the buckets attached to the rim of a large-diameter wheel. The action of recycling water is the same in all such accessories. This one is connected to a contemporary Fleischmann vertical steam engine. (140 by 90 mm; height 140 mm.)

young boys. However, the water caused the tinplate to rust, causing damage which inevitably led to a lot of these toys being discarded.

Water toys can be divided broadly into two groups: those involving buckets; and fountains and ponds with an inbuilt circulating pump. The first type is more frequently found today. The basic layout of these toys comprises a row of buckets attached to a chain forming an endless loop over two gearwheels. The lower gear was immersed in a water reservoir; when the machine was operated, the wa-

ter would be scooped up, carried over the loop and then dumped back into the reservoir. In a number of these toys the buckets were attached to the circumference of a large wheel, which was immersed in the reservoir, and the water was aimlessly recycled until the operator became bored. In more complex types the water would be diverted by channels and chutes and made to drive another toy, such as a waterwheel or a set of hammers.

In the second group of water toys, the fountains would have a water reservoir of varying size and design, a circulating pump and an often elaborate central fountain head with a complex cascade system. Duck ponds were less complex, having a flock of swans or ducks swimming on the surface of the reservoir. The birds were usually made from tinplate but some were made of celluloid and have become collectors' pieces in their own right. Other variations would have a lighthouse in the centre, or small boats circling a central

This bucket-type water toy was made by Fleischmann about 1950. The well is hand-finished tinplate and the figure is composition. Doll made a similar accessory but without the figure. (130 by 90 mm; height 160 mm.)

Left: *This whimsical toy by Bing, made around 1910, shows a complete mountain scene with a lake, sawmill, chalet and mountain stream. The hidden pump circulates water from the lake to the top of the mountain. It then cascades down the mountain back into the lake and also appears to operate the sawmill. Sawmill scenes were produced by Mohr & Kraus and Marklin. (280 by 250 mm; height 200 mm.)*

Right: *This duck pond was made by Doll around 1920. Ponds tended to be less elaborate than fountains, having a greater surface area to allow swans, ducks, fishes etc to float. The swans shown in this example are made from celluloid and are now collectors' items in their own right. (150 by 150 mm.)*

pivot. In some of the more whimsical accessories the toy took the form of a mountain with streams and trees, giving the effect of woodland.

Some of the pumps were disguised as stand pumps, each with its attendant figure, or as horse troughs or drinking fountains.

'Mann an der Pumpe' was made by Bing in 1935. The man is in brightly coloured sculptured lithoplate and the standpipe conceals a working water pump. (165 by 75 mm; height 140 mm.)

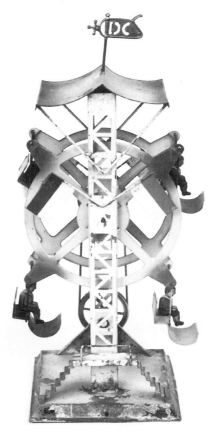

Below: *A magnificent four-seater Ferris wheel by Doll, about 1930. The framework is hand-finished and the figures are lithographed tinplate representing boys dressed in clothes of the period. A number of other firms such as Marklin and Plank also made Ferris wheels. Much larger six- and even twelve-seater versions were made. (120 by 120 mm; height 300 mm.)*

Above: *A superb and colourful carousel in hand-painted tinplate with lithographed tinplate figures. In this example, possibly by Mohr & Kraus, 1920, the figures are spinning around in waltzers. Other carousels may have them riding in aeroplanes or on gallopers. (Diameter 120 mm; height 235 mm.)*

FAIRGROUND TOYS

Elaborate fairground toys represent the pinnacle of accessory construction, being both highly colourful and fascinating to use.

The most interesting examples are those by Marklin and Doll. Both firms made an extensive range, from small swingboats and chairoplanes to elaborate carousels and Ferris wheels. Those made by Marklin frequently had their own music box, providing a suitable fairground atmosphere. All these rides reflected the heyday of the travelling fairgrounds in the 1920s and 1930s, when the rides had to be easily dismantled and transported.

Enjoying themselves on the rides were figures that were spun and thrown about just as real people would have been. The figures were usually made from tinplate, although some were made from a non-metallic plastic material known as composition. Porcelain was used very rarely by a few makers, such as Heino Becker.

Fairground rides have become so popular with collectors that there are now re-

17

Rare fairground double swingboats by Doll, made around 1930. The framework is hand-finished. The original composition figures are still retained. In operation the overhead crankshaft imparts the necessary swinging motion to the boats suspended below. Single and triple versions were also produced. (135 by 70 mm; height 140 mm.)

production figures on the market, but all these are made of composition, not tinplate or porcelain. This point should be borne in mind when considering an expensive purchase.

WORKING MEN

Working men were well represented by accessories. Any of the machine tools mentioned could have had a workman operator. The most frequently made were men

These figures by Wilesco from 1980 represent the final flowering of accessory production. For economy they are made from the same pressing and they use brightly coloured lithographed tinplate. The tab and slot method of construction is clearly visible. The man at the sawbench should be compared with the early figure by Mohr & Kraus (page 9). Although they are almost eighty years apart their actions have not altered. (110 by 80 mm; height 200 mm.)

In this toy by Bing from about 1925 the figure gives the impression of operating a hurdy-gurdy, which contains a simple musical box. It is made from lithographed tinplate and is very rare and desirable. Hurdy-gurdy operators were also made by Carette. (135 by 70 mm; height 135 mm.)

to operate the grinder and the sawbench.

The figures were usually constructed from tinplate. The more expensive ones were three-dimensional, made and painted by hand. The cheaper toys would have been made from two-dimensional flat tinplate and lithographed. Very rarely they were made from wood, the character being brought to life by the use of a sheet of lithographed paper pasted to the front of the toy.

This sausagemaker by Fleischmann was made in about 1950. The use of plastic for the figure indicates that it is a post-war example. The mechanism is tinplate and the endless stream of sausages is made from wood. This toy, in various forms, had been in production for nearly thirty years, firstly by Doll and then by Fleischmann. (135 by 95 mm; height 90 mm.)

*Mohr &
Kraus made
this tinsmith
of litho-
graphed
tinplate in
about 1910.
The back-
ground
represents an
unlikely
timber-framed
building. In
this example
the work
piece is
missing. (125
by 75 mm;
height 125
mm.)*

One of the most interesting is the sausagemaker. He is seen producing an endless stream of sausages. The original toy was made by Doll in the 1920s with a hand-painted tinplate figure. In the 1930s, to reduce cost, the figure was made in lithographed tinplate. Fleischmann took over manufacture and continued to produce the sausagemaker after the Second World War, but then the figure was made from plastic. The toy was eventually discontinued in about 1950. Thus basically the same toy was made for thirty years and changes in the figure indicate its age.

This endearing cobbler was made by Bing in about 1925 and is a most charming fellow. The detailed lithography even includes a cuckoo clock on the wall. The same background was also used for a workman in a tailor's shop. The arm and head are connected so that when in operation both move together. (140 by 80 mm; height 125 mm.)

20

MISCELLANEOUS TOYS

Other toys range from the simplest rotating toys, such as windmills and fans, to complex items such as working dynamos, and some that are unusual and whimsical.

Windmills were by far the most popular of the simpler toys, most firms producing one or more models. Some were quite accurate copies of full-sized mills; others were more toylike, with their own occupants in gaily coloured lithography.

This musical rarity by Bing, made in about 1920, has three bells which make a marvellous sound when struck in sequence by the hammers. The pitch could be altered by adjusting the screws on each chime, although regrettably the tune could not be varied.

Hans Eberl made this rare tinplate elephant and organ in about 1906. The head of the elephant is pivoted and the trunk connected to the handle of the barrel organ, which contains the musical box. When coupled to a steam engine, the elephant gives the impression of playing the organ.

Before radio and television musical toys were quite common. Bing produced a simple toy with bells that could strike out a simple tune. Many parents must have regretted that purchase! Musical boxes disguised as barrel organs and hurdy-gurdies were also popular. They usually had a tinplate organ grinder, but Hans Eberl produced a toy with an elephant operating the box with its trunk! Doll produced a series of military figures playing musical instruments. They were dressed in the uniform of the period of Frederick the Great in the eighteenth century.

Chromotropes (moving colours) comprised two superimposed circular glasses, brilliantly coloured, one of which rotated in front of the other. There were a number of variations on this theme. Marklin made one with three small rotating discs painted yellow, blue and red. Each was attached to a small gear, which intermeshed with a central gear attached to the drive pulley. When the central gear rotated, the three outer discs were spun. Depending on the speed of rotation, they would produce a kaleidoscope of different patterns.

All the major companies produced a dynamo. These could usually be bought separately and coupled to the engine, or they came as an integral part of a complete steam-driven generating plant. They gave only enough power to light a dolls' house, producing between 1.5 and 4.5 volts at about 4000 rpm, but even then, when the light was switched on, the engine would struggle as they were generally underpowered. Today many do not work because the soft iron magnets have become demagnetised. To re-energise them is a complex procedure and as the appearance of the toy is not affected it is better to leave them in the non-working state.

Butter churners are often seen and most were for effect only. However, a number of firms made larger-scale butter churners that did actually work.

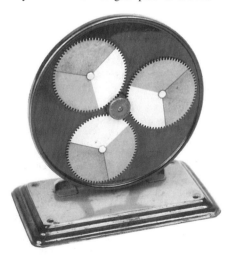

Chromotropes and their variations are very rare toys. In this example by Marklin from about 1925 the three intermeshed discs rotate and produce a variety of different patterns, depending upon the speed of rotation. (130 by 80 mm; height 140 mm; diameter of disc 120 mm.)

Right: *This electricity-generating set was made by Marklin in about 1935. The boiler and engine are set horizontally and there is no integral lamp. It is fired by methylated spirits but another version was made which was electrically heated. It was also available without the dynamo, which could be bought separately. This version produced 2.5 volts. (270 by 270 mm; height 320 mm.)*

Below: *This stand-alone dynamo with an integral lamp was made by the English firm of Bowman (Dereham) in about 1930. The dynamo is coupled to a later steam engine by Bowman (Luton).*

Agricultural toys were only rarely produced, which is a little surprising as Europe was still largely an agricultural society until the First World War. The toys produced were often types of graders. Full-size, they would have been used to grade crops such as potatoes; in miniature they were used to grade sand.

Above: *This delightful butter churner was made by Marklin in about 1930. The lid can be removed and cream placed inside. After rotation for a suitable period butter is produced. Made from hand-finished tinplate with a glass bowl. (140 by 80 mm; height 120 mm.)*

Left: *This hand-finished agricultural accessory was made by Bing in about 1912. In miniature form it can grade sand, whereas the full-size version would have been used to grade crops such as potatoes. The storage hopper at the top of the toy is missing in this example.*

This straw cutter by Bing was made about 1900 for the German market, as can be seen from the German inscription. As the flywheel rotates the straw is pulled through the machine and chopped by two blades attached to the spokes. This is an extremely large and heavy accessory and would require the largest of toy steam engines to operate it. In hand-finished tinplate and diecast parts. (230 by 90 mm; height 170 mm.)

The humble lineshaft was an essential accessory if more than one toy was to be operated at a time. It comprised a central rotating shaft, supported at both ends by columns. The shaft possessed a number of pulleys, one of which was connected to the engine. The remaining pulleys could then be used to drive additional accessories. Today they can serve as a useful source of pulleys as replacements for those missing from more important accessories.

Meccano, originated by Frank Hornby as Mechanics Made Easy, was a construc-tional set making use of metal girders, rods and pulleys, held together with nuts and bolts. A number of British manufacturers, such as Bowman and Hobbies, made their engines and accessories with holes in the baseplates to match those of Meccano. Thus they could be integrated with a Meccano-built model. Also the young engineer could construct his own accessories using Meccano parts. In this way he could produce an endless variety of toys which could be built and rebuilt at will.

Typical examples of lineshafts, by Marklin, 1930. They each possess a number of pulleys to drive accessories.

24

This presentation set by the English firm of SEL, about 1950, contains their major steam engine along with three accessories: a fan, a grindstone and a saw. They are all rather small and can be recognised by the use of plastic for the body. (All about 60 by 40 mm; height 50 mm.)

COLLECTION AND CARE

Collecting accessories was once considered a minority interest by toy collectors. However, the cost of steam engines has steadily increased and accessories are now seen as a more affordable toy. This increase in popularity, however, is offset by their comparative rarity, especially of the more elaborate types such as fairgrounds and fountains.

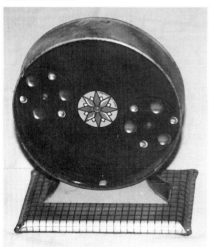

This variation of a chromotrope has glass beads describing a pattern across the face of the toy. It is made from lithographed tinplate. German, about 1930.

Most novice collectors will begin by collecting almost anything. Specialising in a particular area will depend on how much time and money one is willing to invest and whether one wants an interesting and colourful collection.

The accessories most likely to be found are hammers and machine tools, and therefore examples in good condition should be sought. The earlier hand-made and hand-painted accessories, especially those by Marklin, are particularly highly prized. Unfortunately, however, because of the way they were constructed they tend to be in a more distressed condition than later lithographed examples. Nevertheless they still command a premium over them. Fairground accessories can now cost several hundreds or even thousands of pounds, whilst it is not difficult to find hammers or simple water toys for less than £50. Groups with working men are much rarer and are consequently more expensive.

This shoemaker by Plank, about 1900, is an extremely rare piece, being made entirely of wood. The scene, which is highly detailed, depicts the shoemaker stitching the insole on a shoe. It is printed on paper which is then pasted on to the wooden background.

The reverse shows the rather complex operating mechanism. Even the pulleys are made from wood. The action imparts a to and fro motion to the arms, which simulates a stitching movement.

Accessories may be very difficult to date or to attribute to a particular maker. Those of the well-known firms will have their trademarks, but many will simply have 'Made in Germany' or 'Made in England' and many nothing at all. English accessories tended to be basic and solidly built and never possessed a workman. They were limited to machine tools, although Bowman produced a dynamo in the 1930s. All the brightly coloured and elaborate accessories came from Germany.

26

The two acrobats on parallel bars were made by Kraus Mohr in about 1900. The figures rotate around their arms, which are soldered permanently in position. They can be made to rotate in either direction depending on the arrangement of the drive belts. In hand-finished tinplate. (230 by 90 mm; height 120 mm.)

Toy swapmeets are the main source of accessories. They may be advertised locally or in national publications such as *Exchange and Mart* or in specialist publications such as the *Collectors' Gazette*. Look also in car autojumbles, car boot sales, street markets and auctions. Accessories can be found in antique shops but there they tend to be somewhat overpriced.

It may be difficult to assess whether a toy, especially a complex one, is complete and in working order. In these cases reference to a catalogue can be very useful. Missing driving wheels and pulleys can be replaced by one from a redundant lineshaft. Missing gears may have to be specially made, which is neither easy nor cheap.

The condition of paintwork is of paramount importance. It is impossible to touch up lithographed work. Hand-painted items can be touched up or completely repainted, although the original paintwork is preferable, no matter how distressed. Paintwork can be lightly dusted and covered with a fine coating of linseed oil. On no account use any form of abrasive powder to clean the surface or touch trademarks of the transfer variety.

Joints and mechanisms can be cleaned

This hard-working man by Arnold, about 1920, is sawing logs. The figure is in brightly coloured hand-finished tinplate. Arnold made a similar accessory with a lithographed figure. (120 by 80 mm; height 100 mm.)

with light oil and checked for free movements. Small pins and cranks that have become bent can be adjusted with some fine pliers, but if there is the slightest possibility of causing further damage the toy should be left in its non-working state. WD40 is a good protective lubricant but should be kept away from any paintwork.

Post-war accessories may be forty or

Arnold made the scissors grinder in the 1930s. As the figure operates the treadle, the emery-covered wheel produces a shower of sparks. The flat lithographed figure is dressed in the clothes of the period. (175 by 75 mm; height 140 mm.)

fifty years old, pre-war toys up to one hundred years old, so they should always be handled with care and respect, even though they were made to be played with and will survive a surprising amount of rough handling.

Accessories are small enough to be displayed on shelving, preferably with a glass front. They should, however, be kept away from bright sunlight, which may cause the paintwork to fade, and away from moisture so that they will not rust.

Most accessories are capable of being operated without causing any damage. There is nothing more delightful than seeing a Ferris wheel in operation powered by a contemporary steam engine.

The cooper, produced by Doll from about 1930, is a most attractive toy in hand-finished tinplate. As the cooper moves around the barrel he hammers the hoops into position. (130 by 120 mm; height 120 mm.)

This attractive workshop was made by Basil Harley, using commercially available accessories, to illustrate the setup of a toy workshop. It should be compared to the commercially built stations by Carette and Plank (pages 5 and 14). Machine tools are by Wilesco, 1960, and the engine by Doll, about 1930.

MANUFACTURERS

GERMAN MANUFACTURERS

BING (Nuremberg, 1866-1932)

Founded by Ignaz and Adolph Bing in 1866 as Gebruder Bing, the firm did not begin tinplate toy production until 1885. Just after the First World War Ignaz died at the age of seventy-nine and in 1919 Stephan Bing took over, changing the name to Bing Werke. In the late 1920s there were financial problems and in 1932 the firm was taken over by a consortium of toymakers lead by Falk, Kraus and Bub.

Bing, 1913-23.

CARETTE (Nuremberg, 1886-1917)

George Carette, the son of a Parisian photographer who emigrated to Germany, was adopted by the Hopf family in Nuremberg. Here he set up the firm in 1886 and in 1890 he took Paul Josephstal into partnership. By 1911 they were employing some 450 workers and were making toys for Schoenner, Bing and Bassett Lowke. Because the company was French,

the First World War caused problems and the company ceased trading under the name of Carette in 1915. However, it continued until October 1917 under the name of Paul Josephstal.

Carette, 1898-1904.

DOLL (Nuremberg, 1898-1926)

Founded on 19th December 1898 by Peter Doll in partnership with Isaac Sondheim, the firm produced high-quality accessories, but by the end of the First World War it was in a critical financial situation. On 22nd March 1923 a new company emerged, having issued shares to raise capital. The firm prospered until the 1930s but then experienced difficulties because Peter Doll was Jewish. In 1936 the firm was taken over by

Doll, 1898-1938.

29

Fleischmann, who produced accessories under the name of Doll until 1938. Then the name of Doll was dropped completely and from that date all accessories bore the name of Fleischmann.

FALK (Nuremberg, 1896-1935)

This company was started by Joseph Falk, who had been an apprentice to Carette, in 1895. In 1909 he took over some of Schoenner's toys, notably his stork-leg locomotive. His catalogue of 1930 offered 139 steam engines and 155 accessories. However, by 1936 the depression had taken its toll and the firm was taken over by Hans and Fritz Schaller.

Falk, all dates.

FLEISCHMANN (Nuremberg, from 1887)

This firm was founded by Jean Fleischmann in 1887. After his death in 1917 it was run by his wife and brother until 1940, and then by his sons Johan and Emile. In 1928 they acquired Leonhart Staut and 1936 Doll & Company. Then, using the name of Doll, they started to market accessories on a large scale. After the Second World War they continued accessory construction until 1969, when the production of steam toys and accessories ceased. Other toys are still produced.

Fleischmann, 1945 onwards.

HESS (Nuremburg, 1826-1941)

One of the oldest toymaking firms, Hess was started in 1828 but it was not until 1866 that Leonard Joseph Hess, the son of the founder, started to make toys and accessories. By 1929 his son Karl could celebrate one hundred years of production. However, in 1941 the firm closed.

MARKLIN (Göppingen, from 1856)

The founder, Theodore Friedrich Marklin, started making toys in 1859, notably a child's oven. After he died in 1866 his wife carried on the business, but it was not until their two sons, Karl and Eugen, were old enough to help that the firm prospered. In 1891 Ludwig Lutz became a partner and the name was changed to Marklin Brothers & Company. Steam toy and accessory production stopped in 1940 although high-quality toys are still made.

Marklin, 1891-1940.

Marklin, 1930-54.

PLANK (Nuremberg, 1866-1932)

Founded by Ernst Plank in 1866, the firm produced a wide variety of toys such as magic lanterns and optical toys as well as steam accessories. After the death of Ernst the firm was run by his son Karl. During the depression, on 7th September 1932, it was taken over by the Schaller brothers. Toy production ceased but the firm continued to make tin goods until it finally went into liquidation in 1980.

Plank, 1891-1940.

SCHOENNER (Nuremberg, 1875-1917)

Founded in 1875 by Jean Schoenner and Sigmund Shukert, this firm produced the usual range of toys, magic lanterns, steam toys and accessories. By July 1893 they had made 500,000 steam engines. However, in 1909 Falk took over much of its production and in 1917 the firm finally closed down.

Schoenner, all dates.

WILESCO (Wilhelm Schroder, Ludenscheid, from 1912)

The firm was originally founded in 1912 by Wilhelm Schroder and Ernst Wartmann, producing household goods. It was not until 1950 that they started to produce steam toys and accessories. They are the only German toymakers making accessories today.

Wilesco, 1950 on-wards.

OTHER GERMAN MAKERS

Minor German firms that made accessories include:
Karl Arnold (Nuremberg, 1906—)
Hans Eberl (Nuremberg, 1886-99)
Bernard Hommola (1897-1903)
Willy Honsel (1948-54)
Wilhelm Kraus (1932-8)
Kraus Mohr (1895-1903)
Mohr & Kraus (1901-23)

BRITISH MANUFACTURERS
BOWMAN MODELS (Dereham, Norfolk, 1923-35)

Bowman Models was founded by Geoffrey Bowman Jenkins in association with Hobbies Ltd of Dereham. They initially produced steam launches and stationary steam engines and accessories. By

This novelty butcher by Becker of Zschopau, 1910, is shown chopping a pig's head. It was made and painted entirely by hand.

1935 Bowman had given up his association with Hobbies and his place had been taken by Geoffrey Malins.

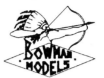

Bowman, all dates.

MAMOD LTD (from 1936)

Mamod is the best-known steam toy-maker in Britain. It was founded by Geoffrey Malins in 1935 when Geoffrey Bowman Jenkins gave up steam toymaking for Hobbies Ltd. The following year he started to produce steam toys under the Mamod trademark. The toys resembled those of Bowman, although they abandoned the old Bowman range and produced their own limited range of accessories. Production was soon halted by the Second World War, after which the firm was supervised by Eric Malins. They ran into difficulties in the 1980s and were taken over by Jedmond Engineers, who, in turn, were taken over by Thomas Johnson Ltd, based in Reading. The firm is still making steam toys and accessories.

Mamod Ltd, all dates.

FURTHER READING

Becker, Carlernst and Hass. *The Other Nurembergers,* volumes 1-6. Frankfurt am Main, 1978-81.
The Collectors' Gazette.
Gordon, Bob. *Toy Steam Engines.* Shire, 1985.
Harley, Basil. *Toyshop Steam.* MAP, 1978.
Kaiser, Wolf, and Becker, Carlernst. *Battenburg Sammler-Kataloge.* Battenburg Verlag, 1983.

REPRODUCTION CATALOGUES
Bing Catalogue (1907).
Gamages Christmas Bazaar (1913). David & Charles, 1974.
The Great Toys of George Carette (1911). New Cavendish Books.
Prijscourant No. 2. Merkelbach & Co. (1925). Amsterdam.

PLACES TO VISIT

GREAT BRITAIN
Chester Toy and Doll Museum, 13A Lower Bridge Street, Chester, Cheshire CH1 1RS. Telephone: 01244 346297.
Museum of Childhood, 1 Castle Street, Beaumaris, Anglesey, Gwynedd LL58 8AP. Telephone: 01248 712498.
Pollock's Toy Museum, 1 Scala Street, London W1P 1LT. Telephone: 0171-636 3452.
The Vintage Toy and Train Museum, First Floor, Field's Department Store, Market Place, Sidmouth, Devon EX10 8LU. Telephone: 01395 515124 extension 208.

EUROPE
Marklin Museum, Holzheimerstrasse 8, Göppingen, Germany.
Musée des Arts decoratifs, 107 Rue de Rivoli, 75001 Paris.
Musée du Jouet, 2 Rue de l'Enclose-de-l'Abbaye, 78300 Poissy.
Musée du Rambolitran, 4 Place Jeanne d'Arc, 78120 Rambouillet.
Spielzeugmuseum, Marienplatz, Munich, Germany.
Spielzeugmuseum, Karlstrasse 13, D-8500 Nuremberg, Germany.
Spielzeugmuseum, Fontungstrasse 15, Zürich, Switzerland.
Stockholms Leksaksmuseum, Stockholm, Sweden

A commercial work station by the American firm of Weeden, from about 1920. It comprises a lineshaft, saw, grindstone, trip hammer and double drop hammers. They are all finished in the rather dull orange and green of the manufacturer. These, apart from a windmill, were the only accessories available from this firm in Britain.